Cool Careers in ENGINEERING

CONTENTS

Ephrahim

Susan

Jan

John

Celeste

Angela

Aprille

Penina

Krishna

Stephen

Jananda

Mark

What Do You Want to Be?

Is being an engineer one of your goals?

The good news is that there are many different paths leading there. The people who work in engineering come from many different backgrounds. But they all have one thing in common—they're problem solvers. Engineers are behind everything from MP3s and DVDs, bridges and biofuels, to skyscrapers and satellites, lasers and lunar rovers, and much more.

It's never too soon to think about what you want to be. You probably have lots of things that you like to do—maybe you like doing experiments or drawing pictures. Maybe you like working with numbers or writing stories.

SALLY RIDE
First American Woman in Space
(1951-2012)

The women and men you're about to meet found their careers by doing what they love. As you read this book and do the activities, think about what you like doing. Then follow your interests and see where they take you. You just might find your career, too.

Reach for the stars!

Sally K Ride

Ephrahim Garcia

Cornell University

"We can learn from nature and incorporate natural features into human-made machines to make them work better," Ephrahim says.

Morphing Planes

Growing up, Ephrahim Garcia liked learning about space, science, and new technology. "My dad and I would talk about science and technology and what is possible as a civilization," he says. "I was fascinated by how technology went from the Wright brothers' *Flyer* to a 747." What's next? Ephrahim is working on the answer to that question—airplanes that can morph, or change shape, in midair.

Nature's Solutions

"Nobody has changed the construction of an airplane much in 100 years," Ephrahim says. "Basically, they are fixed-frame structures." He's trying to learn the fundamentals of shape-changing by testing a 9-foot model in a wind tunnel. He's also studying how birds change their shapes. "Sometimes nature's solutions are better than human-made ones," he says.

Extreme Makeover

While working for the military, Ephrahim started to think about how useful it would be if planes could change shape. He envisioned planes that could morph their wings to zoom through narrow airspaces. Or they could tuck their tails under their wings to make quick landings like birds do.

NASA is planning a flexible airplane for the future. This is an illustration of the agency's morphing airplane.

An aerospace engineer . . .

designs, builds, and tests aircraft and spacecraft. Ephrahim uses his skills to develop new types of flexible aircraft. Other **aerospace engineers**

- study how the flow of air around a moving aircraft affects its flight.
- identify and fix problems in the design of an airplane.
- design and build rockets, fighter jets, and helicopters.

Engineering Design Process

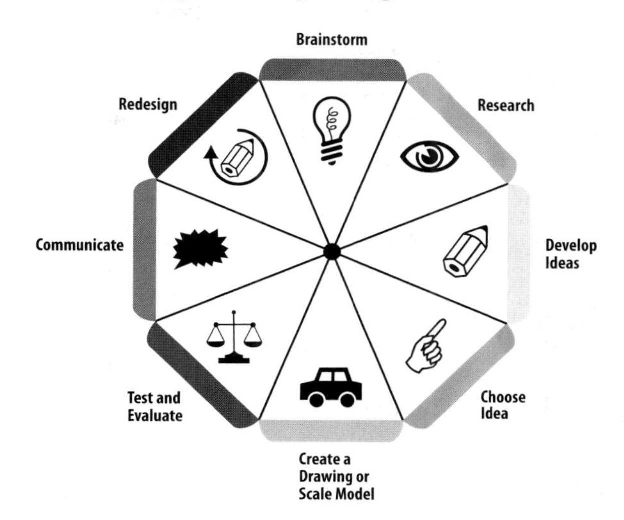

By Design

Nature has inspired many designs. Which inventions do you think were inspired by a

- bird in flight with wings spread?
- bird's beak slightly opened?
- cat's whiskers?
- bur in an animal's fur?

Brainstorm with a partner and then share your ideas with the class.

Go for the Goal

Ephrahim loved science classes in high school. But he didn't know how to become a scientist until a physics teacher recommended that he study engineering in college. What classes do you love? Make a list and write down some of your future goals.

Mother Nature, Engineer

With a team, discuss what you have seen in nature that could be the model for a new technology. Use the *Engineering Design Process* above to come up with a new invention.

- Decide on a goal. What do you see as challenges?
- *Brainstorm*, *Research*, and *Develop* ideas. *Choose* one.
- *Create* a sketch, label its features, and explain why they are important.

Does your drawing *Communicate* your idea? Ask other teams to *Evaluate* it. Use their feedback to guide your *Redesign*.

Check out your answers on page 36.

Susan Helms

United States Air Force

Flying High

Susan Helms wanted to join the Air Force from the time she was little. But her eyesight wasn't good enough to be a pilot. Next choice? Aerospace engineering. While working on the missiles that are attached to fighter jets, she thought, "There must be a way that an engineer can fly." Well, there was. Susan trained as a flight-test engineer and designed tests to study the way airplanes fly in the sky. Finally, she got to fly, sitting behind the pilot. Susan has flown in more than 30 types of military aircraft as the pilot pushed the machines to their limits, screaming across the sky. "I don't remember the scary moments," Susan says. "I just remember the thrill of having a good time and doing a good job at the same time."

Saluting Susan

Since being commissioned as an officer in the U.S. Air Force, Susan has climbed the ranks and currently is Lieutenant General Helms! She helps keep our country safe by directing military planning for space, cyberspace, and deterrence for U.S. Strategic Command—made up of women and men from all four branches of the military.

Above and Beyond

Then Susan really took off—into space! She became an astronaut and flew on four Space Shuttle missions. The highlight of her career was when she lived on the space station for six months as part of its second crew.

Susan set the record for the longest space walk ever—nearly 9 hours.

An Air Force lieutenant general . . .

is a three-star general, nominated by the president, who serves as a high-level officer at various command centers and the Pentagon. Susan leads people responsible for many space-related operations. Other **Air Force lieutenant generals**

- develop next-generation aircraft.
- manage communication and surveillance satellites.
- ensure troops are protected and prepared.

Air Force

What affects an airplane's distance and time in the air? It's the amount of thrust, or force, it receives to propel it forward.

Make two paper airplanes of the same size and shape. Hold one in your right hand and one in your left. Predict which plane will fly farther. Now gently launch both planes at the same time and in the same direction three times. Have a partner measure and log the distances flown. Then calculate the average distance for each hand. How did the thrust vary? Which plane flew farther?

Fly Far, Fly Long

Be a flight-test engineer. With a team, make three different kinds of paper airplanes. You can modify the design by changing the size of the body, the shape of the wings, or the weight of the plane. Launch each plane several times. Be sure to record the distance and length of time in the air for each flight. As a class, compare your results. Discuss which features make a plane fly farther or longer, and why.

Think About

"I was creative as a kid and took a lot of art lessons," Susan says. "The neat thing about engineering is that it's kind of the ultimate combination of mathematics and creativity." Susan used her skills to draw airplane designs in college.

Choose a subject and write a paragraph that explains how you could combine your creativity with that subject to create a career you would enjoy.

Jan Talbot

University of California, San Diego

"When you have a passion for something, you don't listen to people who say you can't."

Always a Challenge

In college, Jan Talbot was in classes with hundreds of other students—and sometimes she was the only woman. When the professor would say, "Good morning, gentlemen," she had to wave her hand to remind him she was there, too! She says she never got discouraged. Jan decided to focus on chemistry because it was her worst subject. That way, Jan laughs, she would always have a good challenge to enjoy.

Variety Is the Spice of Life

Jan combines chemistry, physics, math, and engineering in her work. "I like the variety and the ability to use all those tools to solve a problem," she says. Jan does lots of experiments. She uses chemistry to create better materials, including ones used to make the flat-panel screens that are popular in today's computers and televisions. Another growing field is nanotechnology—making incredibly small things, atom by atom or molecule by molecule.

Jan enjoys a challenge such as a hike in the desert.

A chemical engineer . . .

uses chemistry to find new, different, and better ways to do lots of things that affect the way we live. Jan creates new materials for TVs and computers. Other **chemical engineers**

- clean up polluted drinking water.
- develop flavors for foods and drinks.
- create fireproof fabrics.
- invent paints that are environmentally friendly.

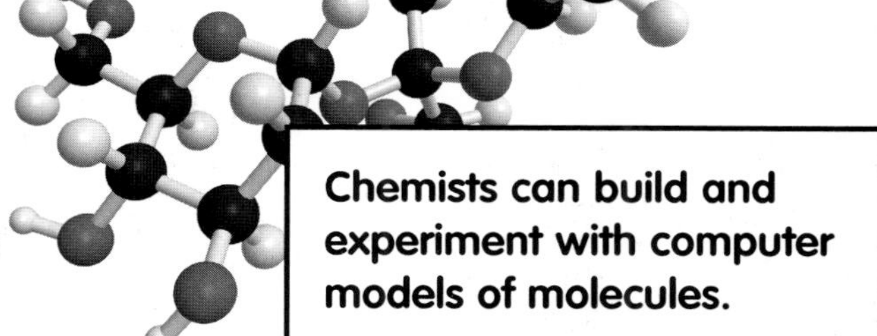

Chemists can build and experiment with computer models of molecules.

Chemical Recipes

Chemists use shorthand or chemical formulas to describe the elements that make up a chemical compound. On a sheet of paper, write either the name of the chemical compound or the chemical formula.

- $C_6H_{12}O_6$
- $NaHCO_3$
- Water
- HCl
- Table salt
- Carbon dioxide

From Bad to Best

Chemistry was Jan's worst subject. Yet she mastered it because she enjoyed the challenge. Which subject is your biggest challenge? Write down a list of steps you could take to turn it into one of your best subjects. Then stick to it. Revisit your plan periodically. See how you've grown and changed or what steps you might need to adjust.

How Small Is Small?

Nano means small, r-e-a-l-l-y small. There are 10,000,000 nanometers in one centimeter (25,400,000 per inch!). Calculate how many nanometers make up the length of a

- human hair (0.008 cm).
- ladybug (0.635 cm).
- computer key (1.25 cm).

Check out your answers on page 36.

JOHN SELKER

Oregon State University

One Looooooooong Thermometer

John Selker is a different kind of cable guy. He uses cable as a thermometer to monitor the health of ecosystems. John frequently uncoils thousands of meters of fiber-optic cable. He lays the cables across frozen Swiss glaciers, inside waterlogged Czech mines, and through drafty Oregon valleys. It's the same thin-as-a-hair glass fiber that carries cable TV signals. But John uses the spindly strands to do a completely different job—check ice, water, and air temperatures. "My work takes me all over—to different parts of the world, to different fields of science," John says.

Check, Check, Check

John's high-tech thermometer accurately measures temperatures every few seconds, every meter along its length. In just an hour, John can collect tens of thousands of measurements over long distances. In the Pacific Northwest, that gives a truer picture of river temperatures. It can also help farmers and fish share scarce water resources. If farmers use too much water, river temperatures rise—threatening fish such as Chinook salmon. Super-long thermometers can reveal when river temperatures are cool enough to allow farmers to irrigate their crops—without endangering fish.

Treetop Tater Tots

In Oregon's Cascade Mountains, John wanted to measure valley air temperatures. The thick forest made it hard to string up the cable. John's low-tech solution? A potato cannon! He hooked cable to the potato and used the spud gun to thread his airborne thermometer through the branches. Flying taters to the rescue!

John loads a potato cannon. Shooting spuds is his creative solution to an engineering challenge.

An ecological engineer . . .

combines ecology and engineering to monitor, design, and restore ecosystems to help both people and the environment. John measures temperatures to understand their environmental impacts. Other **ecological engineers**

- safely remove dams to restore free-flowing rivers.
- design wetlands that naturally treat storm water and waste.
- create ways to block the spread of invasive species.
- design tools to investigate ecosystems.

Nosing Around

As a young boy, John knew he wanted to do something to help the world. But what? He let his interest in science and nature guide him.

After studying physics in college, John worked in Africa and Asia, then studied agricultural engineering. "Follow your nose. You'll find the world has lots of questions waiting to be answered," John says.

What are your interests right now? If you follow them, where do you think they'll lead you?

Balance the Scales

The Fahrenheit temperature scale (°F) is used by most people in the U.S. But engineers such as John measure temperature using the Celsius scale (°C). Convert the temperatures on the chart to degrees Fahrenheit using this formula

$$°F = (°C \times 9/5) + 32$$

Then answer these fishy questions.

- For which fish species is 68°F the highest ideal temperature?
- Which fish species thrives at temperatures from 39°F to 50°F?

On a sheet of paper, draw a thermometer that shows °C on the left and °F on the right. Label the boiling and freezing points of water for each scale. Then label the temperature range for each fish species.

Ideal Temperatures

Species	Temperature Range (°C)
Lake trout	4–10
King salmon	11–12
Albacore tuna (adults)	14–20

Check out your answers on page 36.

Celeste Ford

Stellar Solutions

Taking Off

When Celeste Ford took her first aerospace engineering class, she knew she was onto something. "It was so much fun. For many years in math and science classes, we learned the basics but didn't apply it to anything," she says. "All of a sudden it came to life. We were actually building model airplanes and making them work."

On Her Own

For years, Celeste worked on satellite TV and spy satellites for other companies. Then she started her own—Stellar Solutions. "Our company makes sure that everything works together," she says. One project, a network of seismic ground sensors and their own satellite called *QuakeSat I*, launched in 2003. "We hope that someday you'll be able to turn on your TV and see earthquake forecasts just like you do with hurricanes."

Bring It On

Running her own company, Celeste gets to pick and choose what she works on. "We like to solve the hard problems," she says. Celeste regrets not starting her company sooner. "Everyone should consider it because it's the best way to focus on what you like to do and what you're good at."

Celeste is holding a model of *QuakeSat I*.

An entrepreneur . . .

starts and runs businesses. Celeste started her own company after she had worked for other people, gaining experience and knowledge. There is no typical activity that **entrepreneurs** are involved in, but they can

- develop new products or services.
- think up how to make old things in new ways.
- take over and change existing companies.

I'd Like to Redesign

Celeste is not only an entrepreneur, she's an engineer. That means she's always thinking of ways to improve things. If you were an engineer, what would you like to improve and why? Is it your clock radio, cell phone, or toaster? Or is it some piece of sports equipment?

Draw detailed *before* and *after* illustrations of your redesign. Include captions that describe what each new feature does and why it's important.

In Good Company

Celeste says that starting your own company is "the best way to focus on what you like to do and what you're good at." It takes leadership, determination, confidence, and good communication skills.

Write a paragraph about the type of company you would like to start. How will you be able to use what you're good at? Now present your business idea to your class and look for employees among your classmates.

About You

It takes courage to be an entrepreneur. Celeste is not afraid to take risks and to fail. She knows it's all part of the process of being successful. You need to try and try again until things work! Write a paragraph about something you have done that involved courage, taking a risk, and the possibility of failure. What did you learn about yourself from it?

Angela Belcher

Massachusetts Institute of Technology

Seashore Inspiration

What can an engineer learn from a seashell? In college, Angela Belcher studied the abalone shellfish. "The organism has learned how to make a shell better than we can as materials scientists," she says. Imagine a flawless brick wall so small the bricks are individual molecules—a wall that builds itself and is completely recyclable. Now you get the picture.

A Factory in a Beaker?

What if you could persuade nature to build things in any shape, out of anything, one delicate atom at a time? That's the goal of Angela's lab. The trick is getting tiny biological parts, such as viruses and bacteria, to work with nonbiological parts, such as the atoms used to build computer chips. Already, Angela has coaxed harmless bacteria into building rechargeable batteries—the kind used to power electric cars and laptop computers.

Room for Growth

Angela also plans materials that can diagnose cancer, store vaccines, replace bones, and form lighter cars. "A lot of it is unraveling the secrets of nature and then applying what you learn in a way that can benefit society," she says. Next time you're at the beach, take notes. You might unravel a secret or two.

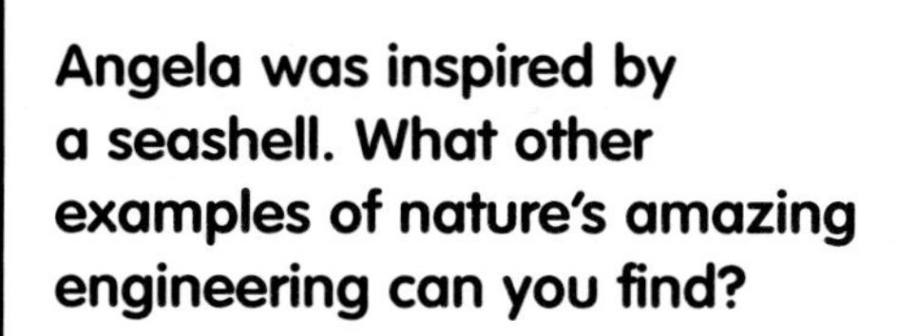

Angela was inspired by a seashell. What other examples of nature's amazing engineering can you find?

A materials engineer . . .

creates new materials with new features or improves existing ones. Angela uses nature to produce new materials. Other **materials engineers**

- improve sports equipment.
- design fabrics that are waterproof, fireproof, or bulletproof.
- find ways to recycle old materials.
- purify metals removed from mines.

Bees are some of nature's tiniest engineers.

Born Engineers

Angela learns a lot from nature. What can you learn from one of nature's amazing engineers? With a partner, research one of the questions below. Then create a science poster. Be sure to title your poster and write a description of what the creature builds and how the creature builds it. Also draw a detailed illustration of the little engineer at work.

- How do hummingbirds build their nests?
- How do spiders make a web?
- How do geckos climb walls?
- How do honeybees make honeycomb?

As a class, take turns sharing your posters and what you've learned.

Nature's Poetry

A seashell inspired Angela. What part of nature inspires you? It could be something you've seen at a zoo, aquarium, botanical garden, or even in your own neighborhood. Write a poem that describes what it is and how it inspires you.

Is It 4 U?

As an engineer, Angela enjoys

- unraveling the secrets of nature.
- working with materials smaller than the width of a soap bubble.
- creating new materials.
- applying what she knows to help others.

What parts of Angela's job would you like? Write a paragraph that explains why you think you'd make a good engineer.

APRILLE ERICSSON

NASA Goddard Space Flight Center

Help from Far Away

How do satellites that orbit hundreds of kilometers above Earth help us? Ask Aprille Ericsson, who puts some of those satellites into orbit. She's worked on spacecraft that study tropical rainfall, the origins of the Universe, and the effects of solar flares on our planet. "The hardware that I build produces scientific data that allow us to understand the Universe and our environment better and help us in our daily lives and communities," she says.

Puzzling and Questioning

Aprille grew up in the projects in Brooklyn. There her mom and grandfather, both engineers, prodded her to use her mind—and her hands. "Picking apart and looking at things around you, asking questions, and then trying to put the puzzles together were all important skills that I learned as a kid," she says. "I've been able to apply them all in my career."

Big to Small

At Howard University, Aprille studied how big structures, such as space stations, flex and vibrate in orbit. It's like architecture, but for stuff that moves. Then she wrote software programs to control the orbits of smaller satellites. Now she manages the design of the instruments, such as X-ray cameras, that go on these cosmic cruisers.

Aprille reviews the construction and testing of a NASA satellite.

A mechanical engineer . . .

designs, builds, and tests machines and structures. Aprille manages the teams of scientists and engineers who build instruments for satellites. Other **mechanical engineers**

- build rovers that explore planets.
- make cars safer in crashes.
- test models of bridges and skyscrapers.
- use computer-aided design (CAD) software.

Paper and Pennies

With a team, imagine you're a mechanical engineer. Work together to design and build a model bridge, and then test it.

- Create a design for a bridge that will span 18 centimeters (7 inches).
- Use no more than eight sheets of paper.
- Fold, roll, twist, crumple, or change the shape of the paper in any way you wish.

What's the catch? There are no scissors allowed, only tape. Add pennies one at a time to test the strength of your bridge. How many pennies will each bridge support? Which design is most effective? Why? Which design is least effective, and why? How will you redesign your bridge to support more pennies?

Egg-citing

With a team, design and construct a "spacecraft" to protect this "instrument," an egg, from a fall. Each team gets one raw egg. Here are some ideas for materials to use to build your spacecraft.

- Fabric
- Paper or plastic bags
- Balloons
- Straws
- Craft sticks
- Paper
- String
- Masking tape
- Paper clips

Brainstorm ideas for a spacecraft design. Sketch your design, and then get to work! Build your spacecraft and secure your egg inside it. Ask your teacher to test-drop each spacecraft from a one-story window. Which spacecraft protected the egg? Why? Which spacecraft failed? Why? As a class, discuss why engineers test spacecraft many times before sending one into space.

Penina "Penny" Axelrad

University of Colorado at Boulder

Think Globally

As a young girl, Penny Axelrad thought she wanted to become an astronaut. Then she found an exciting way to work in space without going there. She started working with the Global Positioning System (GPS) to track the space station. Penny was hooked! "GPS kept changing, and I kept learning new things, so I've stayed in this field for more than 20 years now," she says.

What's GPS?

GPS is a worldwide navigation system made up of satellites orbiting our planet. They send radio signals to receivers on the ground and can pinpoint any place on Earth. If your family's car has a GPS receiver, the signals can show you where you are on a map on a dashboard computer screen. Penny's job is to find the best way to use GPS navigation satellites to track other satellites, such as those that photograph Earth's surface.

Problem Solver

Penny enjoys teaching at the University of Colorado. "If a student doesn't know how to do something, I help her solve the problem by breaking it down to solvable pieces." Does that mean Penny helps her students navigate their way to answers?

In her spare time, Penny works on engineering projects with high school students.

A navigation engineer . . .

uses technology to determine the exact location of people and objects. Penny designs ways to use GPS to track science satellites. Other **navigation engineers**

- design cell phones that can tell you where you are.
- build handheld computers to guide the blind.
- develop systems to track the migration of animals.

Beam Me Down

Imagine you are on a class field trip. Your teacher has a handheld GPS receiver to make sure your class doesn't get lost. GPS satellites beam down radio signals to the receiver. This allows the receiver to pinpoint where on Earth your class is! But how high up is one of those satellites?

The satellite's signal is traveling at the speed of light, 300,000 kilometers (186,000 miles) per second. So, it takes only a few hundredths of a second to reach the receiver. If the signal takes 0.08 seconds to reach the receiver, how far away is the satellite? Since you know the speed and time, rearrange the formula for speed ($s = d/t$) and solve for distance.

Where in the World?

Once GPS pinpoints an object's location, it can display the information using latitude and longitude. These imaginary lines provide an "address" for every place on Earth. Test your navigation skills. With a partner, choose three locations on a map or globe of Earth. On an index card, write down the latitude and longitude coordinates of each, such as 42°N 12°E—that's Rome, Italy. Exchange your card with another team. Who can pinpoint the three locations first? On your mark, get set . . .

Guided Tour

You navigate your way around school every day. Are you a navigation engineer? Write down directions

- from your science classroom to your math classroom.
- from math to the main office.
- from the main office to the media center.

Exchange directions with a partner. Can you follow each other's? Discuss if your directions were clear and accurate. If so, give yourself navigation congratulations! If not, how would you rewrite your directions?

Check out your answers on page 36.

Krishna Shenoy

Stanford University

Brainy Engineer

Reading minds sounds like it belongs in science fiction, not in a science lab. But Krishna Shenoy spends lab time reading the minds of monkeys. He's designed a computer system to intercept the signals from neurons, or nerve cells, in the monkey's brain. The signals are directed to a computer so the monkey can move a computer cursor just by thinking about it!

Power of Thought

Krishna's research on how the brain communicates with muscles may lead to the development of new prosthetics, or artificial body parts. He's working on limbs that can be controlled by people's thoughts. "We're developing technology that can help physicians heal people," he says. "I'm proud of that."

Path to the Brain

In college, Krishna became interested in the brain and then studied electrical engineering. It was the perfect marriage. "We didn't know nearly enough about the brain back then, and we didn't have the technology. We're so far beyond that now." Many people's lives will be touched by today's high-tech tools in the hands of neuroengineers such as Krishna.

Krishna is able to interact with monkeys' brains by using equipment in his lab.

A neuroengineer . . .

designs electronic devices that can interact with the brain and nervous system. Krishna studies how the brain communicates with our arms and legs in hopes of designing thought-controlled prosthetics. Other **neuroengineers**

- design cochlear implants, tiny devices placed inside the ear that help deaf people hear again.
- create pacemakers to help keep hearts beating.
- research new devices that work with the nervous system to control pain.

It's All in Your Head

With a small group, research the brain and make an accurate model of one with clay or another material. Use a different color of clay for each major part of the brain. Then discuss which part of the brain Krishna's technology would focus on if someone had a problem with

- movement, behavior, or memory.
- language, sensations, or recognizing right from left.
- balance, coordination, or fine muscle control.
- breathing, swallowing, or heartbeat.
- hearing and smell.
- vision.

Brainercise

Physical activity improves blood flow to the brain. This can help you concentrate better in school. Think of an exercise you can do—jump rope, sit-ups, or push-ups. Do this exercise for 1 minute and record the date and number of repetitions you complete. Repeat this activity three times a week for several weeks. Then make a line graph of your progress by plotting the number of repetitions against time.

Making Connections

How intricate are nerve cell connections in your brain? Draw 10 dots evenly spaced down the left and right sides of a piece of paper. Each dot represents a nerve cell or neuron. Draw a straight line from the first dot on the left to all 10 dots on the right. Using different colors, repeat this by connecting each dot on the left with the 10 dots on the right.

What a tangled mess! And this doesn't come close to what it's really like. Each neuron might make thousands of connections with other neurons.

"It's not about the physical act of creating, although that's interesting. It's about exploring and understanding."

Stephen Jacobsen

University of Utah

Tinkering for a Living

As a young boy, Stephen Jacobsen loved exploring in his father's workshop. He was more excited about taking apart a phone or clock than sitting in a classroom. Stephen was never head of the class when he was young. But today he heads up a team that creates super-robots.

Power Suit

One of Stephen's projects is a robotic suit that gives its wearer superhuman strength, speed, and endurance. You step into the aluminum boots, pull on the shoulder straps, and grab the hand sensors at the end of its metal arms. "It magnifies the force of your movements 20 times!" Stephen says. Suddenly you can lift a flat-screen TV as if it were an iPod. How? The suit's computer feels your movements because it constantly receives readings from sensors at your feet, hands, and back. Then, the mechanical limbs move with you, doing most of your work. Stephen's robotic suit won't stop a bullet or help you fly. But his work is making comic book supermen and superwomen more realistic every day.

He's an engineering superhero—Stephen operates the robotic suit he created.

Robo-Rex

Stephen and his team have designed and built hundreds of robots. Thousands of amputees wear the robotic artificial arm he designed. Plus, everyone who visits "Jurassic Park – The Ride" at Universal Studios theme park can thank him for creating the giant dinosaurs.

A robotics engineer . . .

designs, builds, and maintains robots—intelligent machines programmed to do all sorts of things. Stephen creates robots for the entertainment, medical, and military fields. Other **robotics engineers**

- design rovers to roam other planets.
- build fast and accurate robotic systems to do manufacturing jobs.
- create robots to collect scientific data in the deep sea or inside volcanoes.
- work on robots that perform delicate and precise surgeries.

Thinking Like an Engineer

If you were a robotics engineer, what kind of robot would you design? What would you want it to do? Would it be used for exploration, entertainment, medicine, manufacturing, or something else? Make a detailed drawing of your robot. Label its features and explain why each is important. Then write a story about your robot. What does it do? Whose lives does it affect?

How Interested R U?

Stephen's advice is, "Follow your own nose. Go where your interests take you." What are some of your interests? Make a list of them. Then choose one and write a paragraph that describes where it could take you.

Tinker Time

As a teenager, Stephen loved to tinker and take things apart. That's engineering in reverse! With a team, investigate the inner workings of something through reverse engineering. Does your teacher or a classmate have a broken electrical appliance, gadget, or gizmo?

- Ask your teacher to cut off all cords and remove all batteries.
- Make a prediction about what you'll find inside.
- Put on safety goggles, open it up, and make a detailed drawing.

Does the inside look the way you thought it would? Now take it apart and try to figure out how it works. Make notes about how the parts fit together and what their purpose might be. Then try to put it back together again. When you are finished, share what you've found with your class.

Jananda Hill

Northrop Grumman

"I've learned a lot about how spacecraft work. I'm becoming somewhat of a rocket scientist."

No Do-overs!

In space, you don't get a second chance. If you've built an advanced satellite, you'd better make sure before launch that it works right. There's no 1-800-REPAIRS up there. That's why Jananda Hill's job—designing software used to test circuit boards for spacecraft—is so crucial to mission success.

The Little Circuit Board That Could

Space is a harsh place, so these are not your average computer circuit boards. To make sure they're space-worthy, engineers must shake, bake, freeze, and squeeze these boards. While that's happening, Jananda's software looks for glitches. Engineers even run the circuit boards in vacuum chambers and zap them with radiation. They try to think of every crazy thing that could happen and see if the little circuit board can survive.

Fun for Our Benefit

Some of the boards will help analyze our environment from orbit, some will gaze into space, and—shhhh!—some are strictly top secret. "It's a lot of fun," Jananda says. "When you see a project you're working on show up on CNN, it makes you feel like you're really doing something important that's contributing to the scientific community, to the nation, and to the world."

A circuit board holds electronic parts and wires together. Check out the one Jananda is holding.

A software engineer . . .

writes programs for computers. Jananda designs software that tests circuit boards for satellites and space probes. Other **software engineers**

- create programming languages to control robots.
- make computer graphics more realistic.
- help you download music faster.
- think of ways to fit more features into a cell phone.

Engineers and astronauts inspect a circuit board.

Wapwn Box Wytx

Huh? The title is a cryptogram—a type of puzzle. Here's how it works. On a sheet of paper, write the alphabet and assign a different letter to each of the 26 letters. Use your code to write a cryptogram like this example.

Mary had a little lamb

XTNB LTZ T SCAASQ STXV

Exchange your cryptogram with a classmate. Try to decode her or his puzzle by looking for frequently used letters, letter patterns, and word spacing. Decoding the letter *E* is a good place to start. Now look at the title again. Can you *Crack the Code*?

Speaking in Code

Software programs are written in computer code. The software that runs your home computer contains tens of millions of lines of code. Engineers use a binary code—1s and 0s—to write computer programs. An eight-digit binary string is used to represent each letter or number. For example, the binary code for the letter *A* is 01000001 and the letter *B* is 01000010. Now create your own eight-digit binary code and write a one-sentence message to exchange with a partner. Be sure to include a key so that your partner can decode your message.

Think About

Jananda is proud of her work because she knows she's making a difference. What makes you feel proud? Write a brief essay that describes something you have done and why you feel good about your accomplishment.

Mark Dunzo

Kimley-Horn

Sensing a Problem

When Mark Dunzo gets stuck in traffic, instead of getting frustrated, he thinks about ways to get the cars moving. Mark designs Intelligent Transportation Systems (ITS) that keep track of the flow of cars and can change the timing of lights. "Sensors can tell if the traffic flow is heavy, like after a ball game," Mark says. "Then the sensors can communicate back to the computer system and say, 'Hey, it's time to use the ball-game timing,' and they keep certain lights green longer."

How It Works

The sensors Mark uses consist of loops of wire that are placed under the road in front of traffic lights. When a car passes over the wires, the sensor sends a message to a central computer. Thanks to Mark's ITS, the computer can coordinate traffic all around the city.

On the Road

As a young boy, Mark wanted to be an engineer like his dad. Then in college he realized that instead of building things, what he really loved were cars, buses, trains, and highways. Suddenly he was on the road to a career.

Mark came by his interest in traffic naturally—he grew up in Los Angeles.

A transportation engineer . . .

works on all areas of transportation, including planning how people and vehicles get around and designing the buildings for those vehicles. Mark designs smart systems for city traffic lights. Other **engineers** design

- city bike paths that are scenic and safe.
- airports or coordinate the air traffic system.
- freight terminals, where the goods carried by trains are transferred to boats or trucks.

Stop, look, and listen. What happens when you push the "walk" button?

Busy Hallways

You can be a transportation engineer for your school. Are hallways crowded between classes? Are there bottlenecks where hallways intersect?

Put together a team and make a chart. For a few days, log traffic flow at different times of the day. Then create a proposal to ease the traffic jams. You might include traffic rules—such as staying to the right, using alternate routes, or designating up and down staircases. Draw a map to show the new traffic flow. Ask your teacher if you could present your plan to the school administration.

Walk, Don't Walk

When you push the "walk" button at an intersection, have you noticed that the light doesn't immediately turn green? Mark says it's designed that way. Pushing the button does make something change, though. Do you know what it is? Check it out the next time you cross at a traffic light. Then discuss your observations with the class.

Traffic Tee-Hee

Q. What goes through towns, up and over hills, but doesn't move?

A. The road!

About Me

The more you know about yourself, the better you'll be able to plan your future. Start an ***About Me Journal*** *so you can investigate your interests, and scout out your skills and strengths.*

Record the date in your journal. Then copy each of the 15 statements below, and write down your responses. Revisit your journal a few times a year to find out how you've changed and grown.

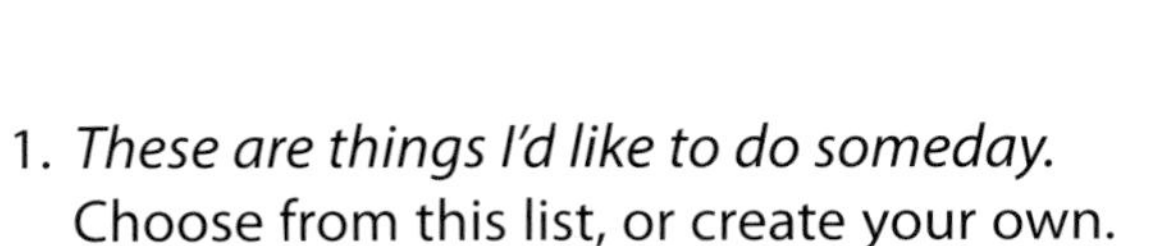

1. *These are things I'd like to do someday.*
   Choose from this list, or create your own.

   - Design new spacecraft
   - Create robotic limbs
   - Start my own company
   - Write computer programs
   - Study Earth from space
   - Analyze data from satellites
   - Design new electronics
   - Research the brain
   - Design computer systems
   - Test aircraft
   - Teach at a university
   - Serve in the military
   - Walk in space
   - Create robots for theme parks
   - Use chemistry to create new materials

2. *These would be part of the perfect job.*
   Choose from this list, or create your own.

   - Working on teams
   - Observing
   - Working outdoors
   - Solving puzzles
   - Building things
   - Brainstorming new ideas
   - Meeting different people
   - Helping people
   - Leading others
   - Traveling

3. *These are things that interest me.*
   Here are some of the interests that people in this book had when they were young. They might inspire some ideas for your journal.
   - Making model airplanes
   - Asking questions
   - Joining the Air Force
   - Taking things apart and putting them back together
   - Building things
   - Figuring out how things work
   - Becoming an engineer
   - Learning about new technology
   - Putting puzzles together
   - Learning about cars, buses, and trains
   - Becoming a pilot
   - Becoming an astronaut
4. *These are my favorite subjects in school.*
5. *These are my favorite places to go on field trips.*
6. *These are things I like to investigate in my free time.*
7. *When I work on teams, I like to do this kind of work.*
8. *When I work alone, I like to do this kind of work.*
9. *These are my strengths—in and out of school.*
10. *These things are important to me—in and out of school.*
11. *These are three activities I like to do.*
12. *These are three activities I don't like to do.*
13. *These are three people I admire.*
14. *If I could invite a special guest to school for the day, this is who I'd choose, and why.*
15. *This is my dream career.*

Careers 4 U!

Which Engineering career is 4 U?

What do you need to do to get there? Do some research and ask some questions. Then, take your ideas about your future—plus inspiration from scientists you've read about—and have a blast mapping out your goals.

On paper or poster board, map your plan. Draw three columns labeled **Middle School**, **High School**, and **College**. Then draw three rows labeled **Classes**, **Electives**, and **Other Activities**. Now, fill in your future.

Don't hold back—reach for the stars!

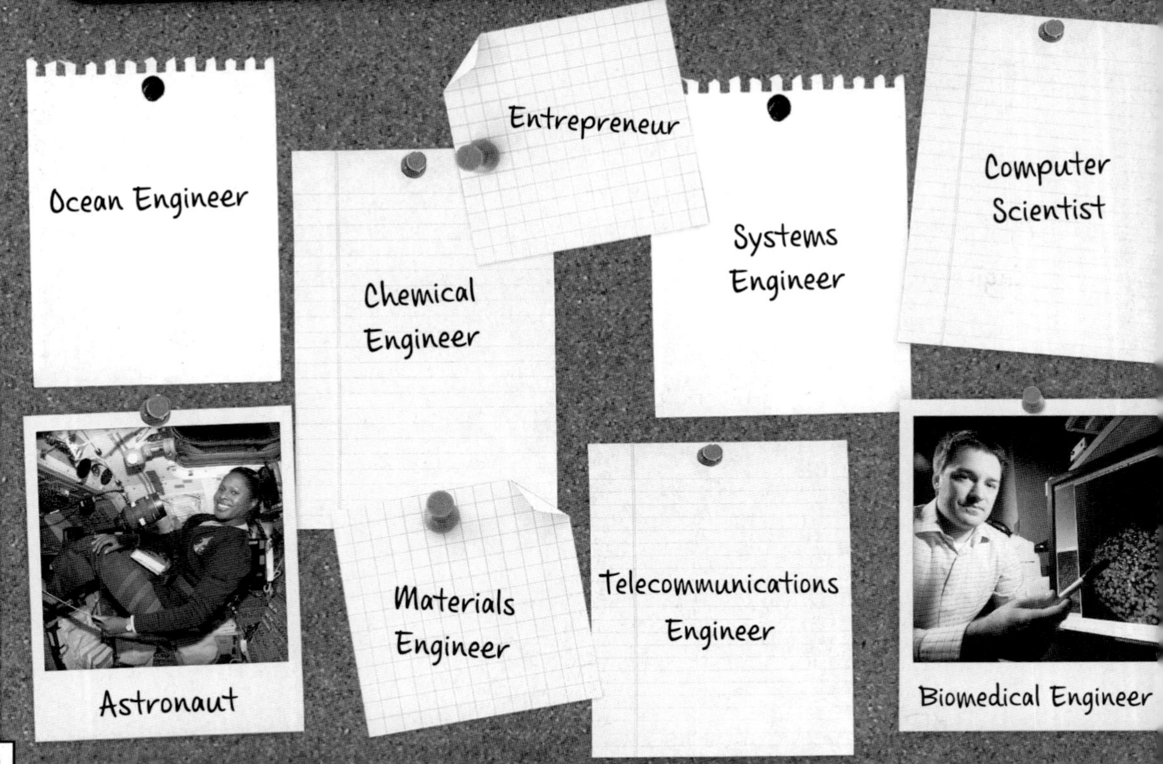

Agricultural Engineer

Electrical Engineer

Technology Teacher

Aerospace Engineer

Neuroengineer

Civil Engineer

Transportation Engineer

Nuclear Engineer

Health and Safety Engineer

Geological Engineer

Renewable Energy Engineer

Ecological Engineer

Navigation Engineer

Coastal Engineer

Mechanical Engineer

Inventor

Software Engineer

Glossary

amputee (n.) A person who has had an arm, leg, or finger removed, usually because it is diseased or damaged. (p. 24)

atom (n.) The smallest particle of an element that keeps its chemical properties. Atoms are the building blocks of all matter. (p. 10)

bacterium (n.) plural **bacteria** A microscopic organism made up of a single cell without a nucleus or other organized cell structures. These organisms occur in three basic shapes—rod, sphere, or spiral. (p. 16)

chemistry (n.) The study of the elements and the ways in which they interact with each other. (pp. 10, 11)

circuit board (n.) A thin rigid board containing an electric circuit; a printed circuit. (pp. 26, 27)

ecology (n.) The study of the relationships between living organisms and the nonliving environment in which they live. (p. 13)

ecosystem (n.) All of the living organisms (plant, animal, and microscopic species) in a given area that interact with each other and their surrounding environment. (pp. 12, 13)

engineering (n.) The application of science, math, and technology to design materials, structures, products, and systems. (pp. 8, 10, 13, 14)

entrepreneur (n.) A person who starts and operates a business. (pp. 14, 15)

manufacturing (n.) Making products from raw materials by hand or by machinery. (p. 25)

molecule (n.) A group of two or more atoms stuck (bonded) tightly together. (pp. 10, 16)

nanotechnology (n.) The technology of building machines, such as electric motors, and eventually whole robots, on a very tiny scale. These devices are typically only a few nanometers in size, and approach the size of individual atoms. (p. 10)

navigate (v.) To plan and direct the route or course of a ship, airplane, or other form of transportation, especially using maps or instruments. (p. 20)

orbit (n.) The path of one body around another, as a result of the force of gravity between them. Examples are a planet's path around the Sun or a moon's path around a planet. (p. 18)

physics (n.) The branch of science that deals with matter, energy, and their interactions. Physics attempts to find laws, usually through math, which accurately describe a wide variety of phenomena throughout the Universe. (pp. 10, 13)

satellite (n.) An object that orbits another object in space. It also refers to something built to orbit the Earth (for example, communications satellites and weather satellites). (pp. 9, 14, 18, 20, 26, 27)

seismic (adj.) Relating to a vibration of Earth caused by an earthquake or something else. (p. 14)

sensor (n.) A device that detects or measures a physical property, such as temperature, which is recorded and can be read by an observer. (pp. 24, 28)

software (n.) A general term for computer programs and files. (pp. 26, 27)

solar flare (n.) An explosion of particles streaming out from a region of the Sun. (p. 18)

surveillance (n.) A close watch kept over someone or something. (p. 9)

virus (n.) A nonliving particle smaller than a bacterium that is able to multiply only inside living cells. Viruses cause many diseases, such as the flu, measles, and the common cold. (p. 16)

Index

CHECK OUT YOUR ANSWERS

AEROSPACE ENGINEER, page 7

By Design

These are actual inventions inspired by nature, but there could be many others, too.

- Airplane wings
- Tweezers
- Sensors
- Velcro

CHEMICAL ENGINEER, page 11

Chemical Recipes

glucose; sodium bicarbonate; H_2O; hydrochloric acid; NaCl; CO_2

How Small Is Small?

human hair

$0.008 \text{ ~~centimeters~~} \times \frac{10{,}000{,}000 \text{ nanometers}}{1 \text{ ~~centimeter~~}}$

= 80,000 nanometers

ladybug

$0.635 \text{ ~~centimeters~~} \times \frac{10{,}000{,}000 \text{ nanometers}}{1 \text{ ~~centimeter~~}}$

= 6,350,000 nanometers

computer key

$1.25 \text{ ~~centimeters~~} \times \frac{10{,}000{,}000 \text{ nanometers}}{1 \text{ ~~centimeter~~}}$

= 12,500,000 nanometers

ECOLOGICAL ENGINEER, page 13

Balance the Scales

Albacore tuna; Lake trout

NAVIGATION ENGINEER, page 21

Beam Me Down

Rearrange $s = \frac{d}{t}$ to $d = s \times t$

Distance of GPS satellite

$\frac{300{,}000 \text{ kilometers}}{1 \text{ ~~second~~}} \times 0.08 \text{ ~~seconds~~} = 24{,}000 \text{ kilometers}$

Distance of GPS satellite

$24{,}000 \text{ ~~kilometers~~} \times \frac{0.62 \text{ miles}}{1 \text{ ~~kilometer~~}} = 14{,}880 \text{ miles}$

IMAGE CREDITS

NASA/Hubble Site/STScI: Cover. iStockphoto.com: pp. 2-5, 30-31 (background). Ephrahim Garcia: p. 2 (Garcia), p. 6 top. U.S. Air Force: p. 2 (Helms), p. 18 top. Jan Talbot: p. 2 (Talbot), p. 10 top, p. 10 bottom. Lina DiGregorio: p. 2 (Selker), p. 12 top, p. 12 bottom. Paolo Vescia: p. 2 (Ford), p. 14 top. Angela Belcher: p. 2 (Belcher). NASA Goddard Space Flight Center: p. 3 (Ericsson), p. 18 top, p. 18 bottom. Larry Harwood: p. 3 (Axelrad), p. 20 top, p. 20 bottom. Dr. Bach-Nga Shenoy: p. 3 (Shenoy), p. 22 top, p. 22 bottom. Sterling Investments LLC: p. 3 (Jacobsen), p. 24 top, p. 24 bottom. Northrop Grumman: p. 3 (Hill), p. 26 top, p. 26 bottom. Kimley-Horn: p. 3 (Dunzo), p. 28 top, p. 28 bottom. Sally Ride Science: p. 4, p. 7, p. 21. NASA: p. 5, p. 27, p. 32 (astronaut), p. 33 (aerospace engineer). Kate Tero/iStockphoto.com: pp. 6-29 (banner). NASA Dryden Flight Research Center: p. 6 bottom. NASA Marshall Space Flight Center: p. 8 bottom. Sanja Gjenero/SXC: p. 9. stanislaff/Shutterstock.com: p. 11. Heather Frackiewicz/Shutterstock.com: p. 13. Stellar Solutions: p. 14 bottom. Slavoljub Pantelic/Shutterstock.com: p. 15. Marsha Miller/UT Austin: p. 16 top. Aguaviva/Shutterstock.com: p. 16 bottom. USDA: p. 17. Trent MacDonald: p. 19. Radu Razvan/Shutterstock.com: p. 23. Matthew Bowden/SXC: p. 25. ilbusca/iStockphoto.com: p. 29. Clara Lam/SXC: p. 30. Gary Meek/Georgia Institute of Technology: p. 32 (biomedical engineer). Jonathan Werner/SXC: pp. 32-33 (corkboard). David Guglielmo/SXC: pp. 32-33 (graph paper). Kelsey Lost/SXC: pp. 32-33 (notebook paper). Sachin Ghodke/SXC: pp. 32-33 (Polaroid). Gelpi/Shutterstock.com: p. 33 (civil engineer). Randy Montoya/Sandia National Laboratories: p. 33 (renewable energy engineer). Clemson University: p. 33 (inventor).

Sally Ride Science is committed to minimizing its environmental impact by using ecologically sound practices. Let's all do our part to create a healthier planet.

These pages are certified as environmentally friendly by the Forest Stewardship Council. They are made with 30% post-consumer waste, bleached without chlorine, and manufactured in the United States using 100% renewable energy.